Disclaimer

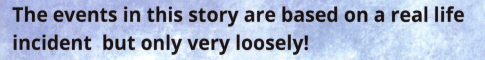

The events in this story are based on a real life incident but only very loosely!

Any resemblance to real persons or actual factual entities and/or cryptids is for the most part coincidental.

Sally Book Bunny lost her key! Where can it be?

Sally said: "I don't know if I left it in that spooky old house -- or if it was taken by a mouse!"

After giving Ziggy & Chippy a snack,
Sally told her friends she would be right back.

Reminding herself that she needed to move her feet, Sally Book Bunny carefully crossed the street...

Thinking the door of the spooky old house would certainly be locked, Sally tried to politely knock (but it swung wide open...)

Sally stepped cautiously into the scary old house... not knowing that right behind her were Ziggy, Chippy and the Mouse!

As their imaginations were running away with them about what they would fear most - suddenly down the stairs floated a gabby green ghost!

Sally pulled a book off the shelf and read aloud (not just to herself): "In this book about myths, we learn that green ghosts don't really exist!"

Upon hearing Sally's proof, the green ghost disappeared. I mean he went *poof*!

Still searching for her missing key, Sally Book Bunny led her friends into another room which had its share of gloom...

This room had many shelves of books from which one could easily choose but just then our friends heard the clattering of ghostly hooves!

Sally read aloud for all to hear (including the fast moving herd of ghostly steer) "Stampedes only happen in the wild west so let me and my friends rest!"

The noise stopped, the dust cleared and the eerie stampede was no more... But just then our friends fell through the floor!

"AAAAAAAAAAAAAAAAHHHHHHHHHHHHHH!!!!!!"

Thankfully our friends landed on a big soft couch —
nobody even said "ouch"!

Ziggy was about to complain about the rumblings in his tummy when suddenly there came a mummy!

Sally thought the mummy looked lonely and needed a friend - so his bandages she offered to mend.

The mummy (who was touched by his new friends' kindness) said: "There are several keys in the parlor — if you get lost, just give a holler."

Following the sounds of an organ, our friends found The Phantom - who being a friendly sort was glad to see them!

Sally had heard the phantom was playing all the wrong keys so she pulled a music book off the shelf and said "try this if you please."

Grateful for Sally's help — The phantom exclaimed (with a flair for the dramatic) — you may find your key in the upstairs attic!

Climbing the stairs with care, our friends saw a man who
wasn't quite all there.

The nearly invisible man said: "You're welcome to take a look-see but please not thru me."

Looking round and round the room, there were many strange things to see but not one of them were Sally's gold key.

Suddenly thru the gloom rode a pretty red-haired witch on a shiny gold broom.

The witch said: "Sally, the key you are searching for is in the lock of that big heavy door."

Sally politely thanked the witch (who began singing beautifully) and then turned to walk toward her key...

As they slowly walked down the long hallway thru time & space, Sally Book Bunny put on a brave face!

Sally and her friends paused at the big heavy door (the gloom was so dense they could hardly see) and then Sally began to slowly turn her glowing key...

"Hey, Sally, time to wake up."

It was all a dream! Sally's "key" (which she never lost) was her love of reading combined with her choice of true-blue friends.

What is <u>your</u> key to happiness?

www.ingramcontent.com/pod-product-compliance
Lightning Source LLC
Chambersburg PA
CBHW041433050326
40690CB00002B/529